W9-BMW-839

Wolf Island

Wolf Island

Celia Godkin

Fitzhenry & Whiteside

To Christopher, Kathryn and Zachary

Wolf Island

Fitzhenry & Whiteside
195 Allstate Parkway
Markham, Ontario L3R 4T8

The author wishes to thank Arnold Skolnick for suggesting the idea of this book.

Designed by Ian Gillen

Typesetting by ISIS Communications

Printed and bound in Singapore

Canadian Cataloguing in Publication Data

Godkin, Celia.
  Wolf Island
ISBN 0-88902-753-6

1. Wolves − Juvenile fiction. I. Title.

PS8563.035W64 1989   jC813'.54   C89-093014-7
PZ10.3.S63Wo 1989

Once there was an island. It was an island with trees and meadows, and many kinds of animals. There were mice, rabbits and deer, squirrels, foxes and several kinds of birds.

All the animals on the island depended on the plants and the other animals for their food and well-being. Some animals ate grass or other plants; some ate insects; some ate other animals. The island animals were healthy. There was plenty of food for all.

Life on the island was peaceful.

A family of wolves lived on the island, too, a male wolf, a female, and their five cubs.

One day the wolf cubs were playing on the beach while their mother and father slept. The cubs found a strange object at the edge of the water.

It was a log raft, nailed together with boards. The cubs had never seen anything like this before. They were very curious. They climbed onto it and sniffed about. Everything smelled different.

While the cubs were poking around, the raft began to drift slowly out into the lake. At first the cubs didn't notice anything wrong. Then, suddenly, there was nothing but water all around the raft.

The cubs were scared. They howled. The mother and father wolf heard the howling and came running down to the water's edge.

They couldn't turn the raft back, and the cubs were too scared to swim, so the adult wolves swam out to the raft and climbed aboard. The raft drifted slowly and steadily over to the mainland. Finally it came to rest on the shore and the wolf family scrambled onto dry land.

There were no longer wolves on the island.

Time passed. Spring grew into summer on the island, and summer into fall. The leaves turned red. Geese flew south, and squirrels stored up nuts for the winter.

Winter was mild that year, with little snow. The green plants were buried under a thin white layer. Deer dug through the snow to find food. They had enough to eat.

Next spring, many fawns were born.

There were now many deer on the island. They were eating large amounts of grass and leaves. The wolf family had kept the deer population down, because wolves eat deer for food. Without wolves to hunt the deer, there were now too many deer on the island for the amount of food available.

Spring grew into summer and summer into fall.
More and more deer ate more and more grass and
more and more leaves.

Rabbits had less to eat, because the deer were eating their food. There were not many baby bunnies born that year.

Foxes had less to eat, because there were fewer
rabbits for them to hunt.

Mice had less to eat, because the deer had eaten the grass and grass seed. There were not many baby mice born that year.

Owls had less to eat, because there were fewer mice for them to hunt. Many animals on the island were hungry.

The first snow fell. Squirrels curled up in their holes, wrapped their tails around them for warmth, and went to sleep. The squirrels were lucky. They had collected a store of nuts for winter.

Other animals did not have winter stores. They had to find food in the snow. Winter is a hard time for animals, but this winter was harder than most. The snow was deep and the weather cold. Most of the plants had already been eaten during the summer and fall. Those few that remained were hard to find, buried deep under the snow.

Rabbits were hungry. Foxes were hungry. Mice were hungry. Owls were hungry. Even the deer were hungry. The whole island was hungry.

The owls flew over to the mainland, looking for mice. They flew over the wolf family walking along the mainland shore. The wolves were thin and hungry, too. They had not found a home, because there were other wolf families on the mainland, and not enough food there for all.

Snow fell for many weeks. The drifts became deeper and deeper. It was harder and harder for animals to find food. Animals grew weaker, and some began to die. The deer were so hungry they gnawed bark from the trees. Trees began to die.

Snow covered the island. The weather grew colder and colder. Ice began to form in the water around the island, and along the mainland coast. It grew thicker and thicker, spreading farther and farther out into the open water. One day there was ice all the way from the mainland to the island.

The wolf family crossed the ice and returned to their old home.

The wolves were hungry when they reached the island, and there were many weak and sick deer for them to eat. The wolves left the healthy deer alone.

Finally, spring came. The snow melted, and grass and leaves began to grow. The wolves remained in their island home, hunting deer. No longer would there be too many deer on the island. Grass and trees would grow again. Rabbits would find enough food. The mice would find enough food. There would be food for the foxes and owls. And there would be food for the deer. The island would have food enough for all.